HOT AND COLD ANIMALS

SUN BEAR OR POLAR BEAR

BY MARILYN EASTON

Children's Press
An imprint of Scholastic Inc.

A special thank you to the team at the Cincinnati Zoo & Botanical Garden for their expert consultation.

Library of Congress Cataloging-in-Publication Data
Names: Easton, Marilyn, author.
Title: Hot and cold animals. Sun bear or Polar bear / by Marilyn Easton.
Other titles: Sun bear or Polar bear
Description: First edition. | New York : Children's Press, an imprint of Scholastic Inc., 2022. | Series: Hot and cold animals | Includes index. | Audience: Ages 5–7. | Audience: Grades K–1. | Summary: "NEW series. Nonfiction, full-color photos and short blocks of text to entertain and explain and how some animals with the same name can survive in very different environments"—Provided by publisher.
Identifiers: LCCN 2021044794 (print) | LCCN 2021044795 (ebook) | ISBN 9781338799361 (library binding) | ISBN 9781338799378 (paperback) | ISBN 9781338799385 (ebk)
Subjects: LCSH: Sun bear—Juvenile literature. | Polar bear—Juvenile literature. | Bears—Juvenile literature. | Habitat (Ecology—Juvenile literature. | BISAC: JUVENILE NONFICTION / Animals / Bears | JUVENILE NONFICTION / Animals / General
Classification: LCC QL737.C27 E222 2022 (print) | LCC QL737.C27 (ebook) | DDC 599.78156—dc23
LC record available at https://lccn.loc.gov/2021044794
LC ebook record available at https://lccn.loc.gov/2021044795

10 9 8 7 6 5 4 3 2 1   22 23 24 25 26

Printed in the U.S.A. 113
First edition, 2022

Book design by Kay Petronio

Photos ©: 5: G&M Therin-Weise/robertharding/age fotostock; 6–7: Valerie/Flickr; 10 bottom left: Mark Newman/Minden Pictures; 10 right: Edwin Giesbers/Minden Pictures; 12–13: Nick Garbutt/Minden Pictures; 16 bottom: Jim McMahon/Mapman ©; 17 left: Jim McMahon/Mapman ©; 18–19: aussieflash/123RF; 22 center: Mark Newman/Getty Images; 23: Fabrice Simon/Biosphoto; 24–25: Alan Porritt/EPA/Shutterstock; 26–27: Klein and Hubert/Minden Pictures; 28 center: ZSSD/Minden Pictures; 29 left: Steven Kazlowski/Minden Pictures; 29 right: Danny Green/Minden Pictures. All other photos © Shutterstock.

SUN BEAR

POLAR BEAR

CONTENTS

MEET THE BEARS

Sun bears and polar bears are very different types of bears. Sun bears live in the warm **rain forest**. They like to climb trees and eat plants and meat.

SUN BEAR

A sun bear's thick fur protects it from twigs and branches. **FACT**

4

Polar bears live in the cold **Arctic**. They like to swim and eat meat.

POLAR BEAR

FACT Polar bears can swim up to 100 miles (161 km) at a time!

COOL FUR

Short fur keeps
the sun bear cool
in the heat.

FACT Sun bears sometimes
sleep in trees.

TREE CLIMBER

These large
paws and strong
claws help
the sun bear
climb trees.

TONGUE-TIED

A sun bear's tongue is 10 inches (25 cm) long!

SO SPECIAL

The shape of every sun bear's golden chest patch is unique.

A sun bear can weigh 60 to 145 pounds (27 to 66 kg).

It has black fur with a patch of golden fur on its chest.

POLAR BEAR CLOSE-UP

A polar bear can weigh 800 to 1,600 pounds (363 to 726 kg).

It has white fur and a black nose.

SUPER SNIFFER
This strong nose can sniff out a seal 20 miles (32 km) away!

CHILLY PAWS
Big webbed paws are for walking on snow and ice.

BLACK AND WHITE

Underneath its white fur, a polar bear has black skin.

IS THAT SNOW?

A polar bear's white fur helps it blend in with the snow and ice.

FACT Polar bear fur is thicker than any other type of bear fur.

SMALL AND BIG

SMALL AND BIG

↑ SUN BEAR

Sun bears are the smallest bear. Polar bears are the largest bear. They look very different!

Their unique
bodies help
them survive in
their **habitats**.
A habitat is
the area where
an animal is
usually found.

POLAR
BEAR

11

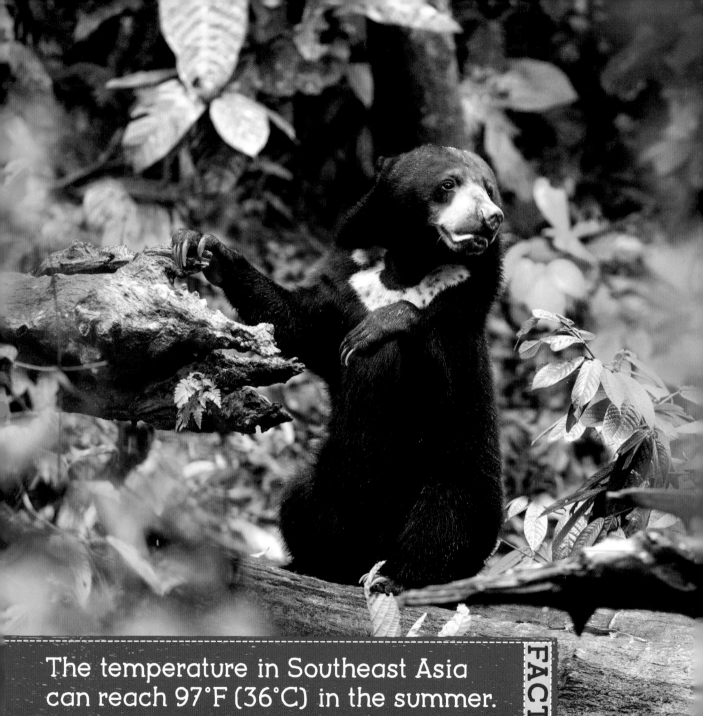

The temperature in Southeast Asia can reach 97°F (36°C) in the summer.

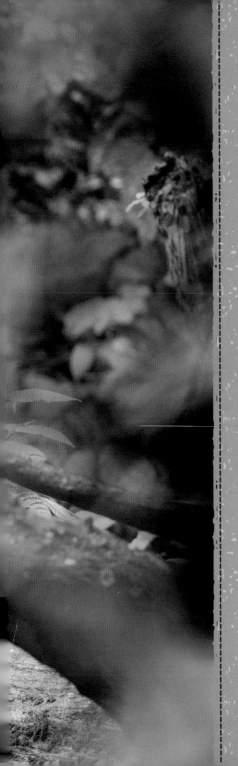

HOT, HOT, HOT!

Sun bears live in the hot, tropical rain forest in Southeast Asia. There are a lot of trees for sun bears to climb.

Even though it is sunny where they live, sun bears don't see too much sun! That's because they are **nocturnal**. A nocturnal animal is active during the night.

HOME, CHILLY HOME

In winter, the Arctic is a very cold place. Polar bears live on frozen seawater called sea ice. When summer comes, the sea ice melts. The polar bears have to move onto land. When winter comes again, the sea ice forms again.

FACT

In winter, the average Arctic temperature is −40°F (−40°C)!

NO NAPPING

Sun bears and polar bears live in very different places. The rain forest has no winter. The Arctic has freezing winters.

Because of where they live, both bears do not **hibernate**.

Arctic Ocean

North America

Europe

Asia

Atlantic Ocean

Pacific Ocean

Africa

South America

Indian Ocean

Pacific Ocean

Australia

Where sun bears live

Southern Ocean

Antarctica

SUN BEARS

Sun bears do not need to hibernate because there is no winter where they live. Polar bears find most of their food in the winter, so they need to stay awake!

POLAR BEARS

Where polar bears live

RUSSIA

FINLAND
SWEDEN

North
+Pole

ARCTIC
OCEAN
(frozen sea)

Greenland
(DENMARK)

NORWAY

Alaska
(U.S.)

ICELAND

ARCTIC CIRCLE

PACIFIC
OCEAN

CANADA

ATLANTIC
OCEAN

UNITED
STATES

N
W   E
S

FACT
Bears that hibernate do not eat or drink for months at a time.

BEEHIVE

Sun bears are also
called honey bears.

FACT

GRAB A BITE

Sun bears are **omnivores**. An omnivore is an animal that eats plants and meat. Sun bears mostly eat bugs, eggs, lizards, fruits, leaves, small **mammals**, and honey. A sun bear uses its long tongue and sharp claws to reach its food. Their claws can tear apart a beehive. Then they slurp the sweet honey inside.

SEAL SNACKS

Polar bears are carnivores. A **carnivore** is an animal that eats meat. Polar bears mostly eat seals. Sometimes they will eat walruses.

A seal comes up for air near some sea ice. A polar bear is waiting at the edge. It grabs the seal and eats a filling meal.

SEAL

FACT Polar bears are the largest carnivores that live on land.

WHAT TO EAT

APPLE

SUN BEAR

Sun bears eat a lot of different things. Polar bears only eat a few things. This may be because the rain forest is full of life and many different food options.

Pythons, Asian leopards, and tigers are some of the sun bear's **predators**.

Polar bears need a lot of food to survive. There are few animals in the Arctic large enough to make a filling meal for a polar bear.

ARCTIC CHAR
FISH

POLAR
BEAR

Polar bears' predators are humans and other polar bears.

Sun bear cubs cannot
see when they are born.

SUN BABIES

Sun bears usually give birth to two babies. Baby bears are called **cubs**. Sun bear cubs live in nests made of leaves. Their nests are inside logs or on the ground. When sun bears are born, they weigh around 11 ounces (312 g). That's about the same weight as two baseballs. They will go out on their own after two years.

CHILLY CUBS

Polar bear mothers give birth to one to three cubs in the winter. The family lives in a den. The playful cubs snuggle with their mom to keep warm.

When spring comes, the mother will show her cubs how to hunt for food. When the cubs are 2–½ years old, they start to take care of themselves.

FACT A newborn polar bear cub is as small as a squirrel.

CUTEST CUBS

Although they grow up to be different adult bears, sun bear cubs and polar bear cubs are the same in many ways. They both stay with their mothers for about two years

SUN BEAR CUB

A mother bear is called a **sow**.

FACT

During this time, their mothers teach them how to survive. The sun bear cubs learn to find food. The polar bear cubs learn to hunt.

POLAR BEAR CUBS

YOU DECIDE!

If you had to choose, would you rather be a sun bear or a polar bear? If you like summer and eating lots of different types of food, you may prefer being a sun bear. If you like winter and don't mind always eating the same food, maybe you would choose to be a polar bear!

FACT

There are eight different types of bears.

GLOSSARY

Arctic (AHRK-tik) – the area around the North Pole

carnivore (KAHR-nuh-vor) – an animal that eats meat

cub (kuhb) – a young animal, such as a bear

habitat (HAB-i-tat) – the place where an animal or a plant is usually found

hibernate (HYE-bur-nate) – to be in a resting state that some animals go through during winter

mammal (MAM-uhl) – a warm-blooded animal that has hair or fur and usually gives birth to live babies

nocturnal (nahk-TUR-nuhl) – active at night

omnivore (AHM-nuh-vor) – an animal that eats both plants and meat

predator (PRED-uh-tur) – an animal that lives by hunting other animals for food

rain forest (RAYN for-ist) – a dense, tropical forest where a lot of rain falls much of the year

sow (sou) – a mother bear

unique (yoo-NEEK) – being the only one of its kind

INDEX

ABOUT THE AUTHOR

Marilyn Easton is the author of more than 50 books. She lives in Los Angeles, California, with her two dogs and one cat. If she had to choose, she would be a sun bear.